GENERAL C
OF EDU

MODEL ANSWERS

APPLIED MATHEMATICS

(Advanced Level)

by
S. SIMONS, PH.D.,
Lecturer in Applied Mathematics,
University of London

PUBLISHED AT
THE ARTEMIS PRESS
Sedgwick Park, Horsham, Sussex
General Editor: M. T. BIZONY, M A, (CANTAB.)

Copyright © 1977, 1973, The Artemis Press Ltd.

All rights reserved. No part of this publication may be transmitted, in any form or by any means, without the written permission of the Publisher.

First published	*July, 1963*
Reprinted	*January, 1965*
Second Edition	*April, 1967*
Third Edition	*January, 1969*
Fourth Edition	*January, 1973*
Revised	*November, 1975*
Fifth Edition	*January, 1977*

ISBN 0 85141 297 1

Printed in England by
Unwin Brothers Limited
The Gresham Press, Old Woking, Surrey

GENERAL ADVICE

The questions in this book are taken from the Advanced Level papers of London University set in recent years. For a note on the structure of the papers please see p. 5.

The best start for 'A'-Level work is to check that you are thoroughly fluent in 'O'-Level material, and revise where necessary. In writing answers to questions, it is well to bear in mind that Applied Mathematics is a mathematical treatment of certain precise physical principles, and that one should therefore state the relevant principle clearly and concisely whenever it is invoked in an answer.

A useful check on the correctness of the form of an equation is to test whether the units of the physical quantities on both sides of the equation are the same; but merely obtaining the right answer to a problem counts for very little. Your work should be set out in such a way that the examiner can follow your line of argument and satisfy himself that it is sound. Corrections should be made neatly so that neither the examiner nor yourself can be in any doubt as to what is meant.

Always begin by selecting those questions which you think you can do with comparative ease. The first part of a question often provides a valuable clue to the best way of dealing with the second part or "rider". Do not allow yourself to get stuck, particularly on one of the short, A-type questions, which carry few marks; you can always come back to a difficult problem later.

It is almost useless to read the model answer to a question unless a serious effort has first been made to solve it without guidance. You should aim at an average of twenty-five minutes for each B-type question and five minutes for each A-type question.

Remember that each mathematical statement should form a grammatical English sentence and should be punctuated accordingly. Correct spelling and orderly presentation of material are important, and neglect of these essentials may cost marks. There is no need to prove or explain statements based entirely on 'O'-Level material. The use of slide-rules is permitted; four-figure mathematical tables and certain formulae will be provided by the University.

CONTENTS

	PAGE
General Advice	3
The Syllabus	5
Section 1: Areas, Volumes and Centroids	7
Section 2: Approximate Integration	10
Section 3: Linear Relations between Two Variables ...	11
Section 4: Vectors...	13
Section 5: Variable Acceleration...	15
Section 6: Vectorial Kinematics	18
Section 7: Laws of Motion	24
Section 8: Collisions and Momentum	30
Section 9: Energy and Power	35
Section 10: Motion of a Particle in a Plane	41
Section 11: Forces in Equilibrium...	47
Section 12: Probability	54
Multiple-Choice Questions	57
Model Test Paper Index	63

The Publishers are indebted to the University of London for permission to reproduce questions from the G.C.E. Examination.

THE SYLLABUS, 1977–1980

Most Examining Boards no longer treat Applied Mathematics as a separate subject. The following is the 'Applied' content of the University of London's combined Syllabus D. The range of topics covered is very similar to the corresponding parts of the syllabuses of other Boards, and questions on it are set in Paper 1 (multiple-choice—see pp. 57–62) and Paper 3. (Paper 2 deals with most of the pure mathematics topics, to which another volume in the G.C.E. Model Answers series is devoted.)

Paper 3 consists of 8 short questions (of which all are to be attempted) and 8 harder questions (of which 4 are to be attempted). Short questions in this book are distinguished by the letter A in the dateline, and carry 4 marks each; harder questions by the letter B, and carry 16 marks each.

Applications of definite integration to calculation of volumes, centroids, centre of mass and mean values. Approximate integration (either the trapezoidal or Simpson's rule may be used). Use of approximations by the first few terms of a Taylor or Maclaurin series. Small increments (one independent variable only). Reduction of the relationship between two variables to a linear form.

Rates of change, e.g. velocity, acceleration, angular velocity. The idea of a vector in two dimensions including addition, subtraction and multiplication by a scalar. Unit vectors. Differentiation of a vector with respect to a scalar variable. Kinematics of a particle moving in a straight line with acceleration either constant or a function of the time, the velocity or the distance, including simple examples of the kinematical differential equations

$$\frac{dv}{dt} = v\frac{dv}{dx} = f(t), f(v) \text{ or } f(x),$$

with solutions.

Displacement, velocity, acceleration, force, momentum as vector quantities. Newton's laws of motion. Dynamics of a particle moving in a straight line. Simple cases of motion of connected particles. The impulse-momentum principle. Conservation of linear momentum. Direct impact of elastic bodies.

Work, energy, power. The work-energy principle. Motion of a particle in two dimensions. Relative velocity. Simple cases of projectiles. Uniform circular motion.

Moments and couples. Simple cases of the equilibrium of particles and rigid bodies under coplanar forces.

Friction. Hooke's Law. The properties of centres of gravity.

Elementary probability; conditional probability; sum and product laws. Understanding of the results

$$P(\bar{A}) = 1 - P(A),$$
$$P(A \cup B) = P(A) + P(B) - P(A \cap B),$$
$$P(A \cap B) = P(B) \cdot P(A|B).$$

The distinction between mutually exclusive and independent events. The meaning and use of simple tree diagrams. Familiarity with simple experiments such as tossing coins, rolling dice or drawing cards.

Section 1 : Areas, Volumes and Centroids

[*The area A under the curve* $y = f(x)$ *between* $x = a$ *and* $x = b$ *is given by*

$$A = \int_a^b f(x)\, dx.$$

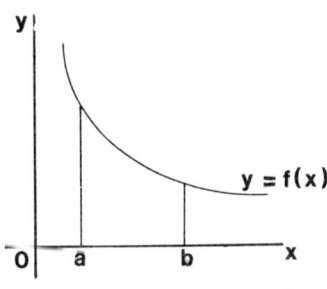

The volume V obtained by rotating this area about the x-axis is given by

$$V = \pi \int_a^b y^2\, dx.$$

The distances of the centroid of A from the x- and y axes, $\bar{y}$ and $\bar{x}$ respectively, are obtained from

$$\bar{y} = \tfrac{1}{2}\int_b^a y^2\, dx / A \quad \text{and} \quad \bar{x} = \int_a^b xy\, dx / A.$$

The centroid of V will lie on the x-axis; its distance from O is

$$\bar{x} = \pi \int_a^b xy^2\, dx / V.$$

The mean value of $f(x)$ between $x = a$ and $x = b$ is given by

$$\bar{f} = \int_a^b f(x)\, dx / (b - a).]$$

Question 1 : (*G.C.E., Specimen Paper for 1977, A.*)

A cap of depth $a/2$ is cut from a solid sphere of radius a. Prove by integration that the volume of the cap is $5\pi a^3/24$.

Answer :

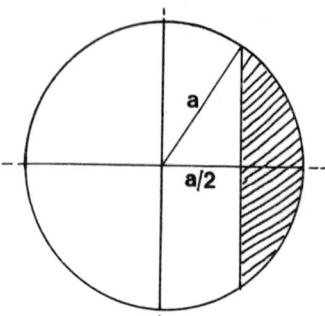

The equation of the circle is $x^2 + y^2 = a^2$, and the cap is obtained by rotating the shaded area in the diagram about the x-axis. The volume of the cap is therefore

$$\pi \int_{a/2}^{a} y^2 \, dx = \pi \int_{a/2}^{a} (a^2 - x^2) \, dx$$

$$= \pi \left[a^2 x - \frac{x^3}{3} \right]_{a/2}^{a} = \pi \left(a^3 - \frac{1}{3}a^3 - \frac{1}{2}a^3 + \frac{1}{24}a^3 \right)$$

$$= 5\pi a^3 / 24, \qquad\qquad Q.E.D.$$

Question 2 : (G.C.E., Specimen Paper for 1977, B.)

(i) The region defined by the inequalities
$$0 \leqslant x \leqslant 2, \quad 0 \leqslant y \leqslant e^x$$
is rotated completely about the x-axis. Find in terms of e the distance from the origin to the centroid of the solid generated.

(ii) Find the mean value of the expression
$$(a + b \sin 2\omega t)^2$$
over the interval $0 \leqslant t \leqslant \pi/\omega$, where a, b, ω are constants.

AREAS, VOLUMES AND CENTROIDS

Answer:

(i) The volume of the disc of thickness dx is $\pi y^2 dx$, and the moment of this about the y-axis is $\pi x y^2 \, dx$. The distance from the origin to the centroid (which by symmetry lies on the x-axis) is thus given by

$$\bar{x} = \frac{\pi \int_0^2 x e^{2x} \, dx}{\pi \int_0^2 e^{2x} \, dx}.$$

Now, integrating by parts,

$$\int_0^2 x e^{2x} \, dx = \left[x \frac{e^{2x}}{2} \right]_0^2 - \frac{1}{2} \int_0^2 e^{2x} \, dx$$

$$= \left[\frac{x}{2} e^{2x} - \frac{e^{2x}}{4} \right]_0^2 = \frac{3}{4} e^4 + \frac{1}{4}.$$

Also $\int_0^2 e^{2x} \, dx = \frac{1}{2} \left[e^{2x} \right]_0^2 = \frac{1}{2}(e^4 - 1).$

Thus $\quad \bar{x} = \dfrac{3e^4 + 1}{4 \times \frac{1}{2}(e^4 - 1)} = \dfrac{3e^4 + 1}{2(e^4 - 1)}.$

(ii) The mean value $\bar{X}$ of $X = (a + b \sin 2\omega t)^2$ for $0 \leqslant t \leqslant \pi/\omega$ is given by

$$\bar{X} = \frac{\int_0^{\pi/\omega} (a + b \sin 2\omega t)^2 \, dt}{\pi/\omega}.$$

To evaluate the integral, let $u = 2\omega t$; then

$$\bar{X} = \frac{\omega}{\pi} \cdot \frac{1}{2\omega} \int_0^{2\pi} (a + b \sin u)^2 \, du$$

$$= (1/2\pi) \int_0^{2\pi} (a^2 + 2ab \sin u + b^2 \sin^2 u) \, du.$$

Now, $\int_0^{2\pi} a^2 \, du = 2\pi a^2$, but $\int_0^{2\pi} 2ab \sin u \, du = 0$ since $\int_0^{2\pi} \sin u \, du = 0$, and $\int_0^{2\pi} b^2 \sin^2 u \, du = \pi b^2$ since $\int_0^{2\pi} \sin^2 u \, du = \pi$. Thus

$$\bar{x} = (1/2\pi)(2\pi a^2 + \pi b^2)$$
$$= (a^2 + \tfrac{1}{2}b^2).$$

Section 2 : Approximate Integration

[THE TRAPEZOIDAL RULE: *To evaluate* $\int_a^b f(x) \, dx$, *divide the integration range a to b into n strips of equal width h. Then, writing* $f_m = f(a + mh)$,

$$\int_a^b f(x) \, dx \approx h(\tfrac{1}{2}f_0 + f_1 + f_2 + \ldots + f_{n-1} + \tfrac{1}{2}f_n).$$

SIMPSON'S RULE: *To evaluate* $\int_a^b f(x) \, dx$, *divide the integration range a to b into an even number, n, of strips all of width h. Then, writing* $f_m = f(a + mh)$,

$$\int_a^b f(x) \, dx \approx \tfrac{1}{3}h\{f_0 + f_n + 2(f_2 + f_4 + \ldots + f_{n-2}) + 4(f_1 + f_3 + \ldots + f_{n-1})\}.]$$

Question 3 : (*G.C.E., Specimen Paper for 1977, B.*)

The speed v m/s of an aircraft accelerating from rest is recorded at 5 second intervals as follows :

t	0	5	10	15	20	25	30
v	0	3·9	10·4	18·1	26·4	35·4	44·7

Use Simpson's rule to estimate the distance in metres covered in the 30 seconds.

Plot a graph of v against t and use this graph to estimate the acceleration, in m/s², of the aircraft at time $t = 20$.

LINEAR RELATION BETWEEN TWO VARIABLES

Answer :

Since $ds/dt = v$ (where $s(t)$ is the distance covered in time t), it follows that
$$s(30) = \int_0^{30} v\, dt.$$
Simpson's Rule is applied here with strip-width $h = 5$, and yields
$$s(30) = (5/3)[0 + 44{\cdot}7 + 2(10{\cdot}4 + 26{\cdot}4) + 4(3{\cdot}9 + 18{\cdot}1 + 35{\cdot}4)] = 579{\cdot}8 \text{ m}.$$

If a is the acceleration, we have $a = dv/dt$, and so the acceleration at any time is the gradient of the tangent to the v-t curve at that point. On the graph, the tangent at $t = 20$ is the line ST, and its gradient is
$$1{\cdot}80 \text{ m s}^{-2},$$
which is therefore the acceleration at $t = 20$.

[*The graph is left as an exercise for the reader. The line ST is the tangent to the curve at the point where $t = 20$, $v = 26{\cdot}4$, and it meets the t-axis at $t = 5{\cdot}3$.*]

Section 3 : Linear Relation between Two Variables

[*If $y = ax + b$, and sets of values (x_1, y_1), (x_2, y_2), ... are given, then a and b can be found by plotting a graph of y against x. The points should lie on a straight line, whose gradient is a and whose intercept on the y-axis is b.*

Certain other relationships between x and y can be reduced to the above linear form. For example,

(i) If $y = ax^n$,   then $\log y = n \log x + \log a$
(ii) If $y = ab^x$,   then $\log y = (\log b)x + \log a$
(iii) If $axy = x - by$,   then $(1/y) = b(1/x) + a$
(iv) If $y = ax + (b/x)$,   then $xy = ax^2 + b$.]

Question 4 : (*G.C.E., Specimen Paper for 1977, B.*)

The table below gives values of the variables u and v, which are related by an equation of the form $v = au + b/u$. Find values of the constants a and b by drawing a straight-line graph relating uv and u^2.

u	1	2	3	4	5
v	12·5	7·0	5·5	5·0	4·9

Answer :

Since $v = au + b/u$, it follows that $uv = au^2 + b$. Thus if a graph is drawn of uv against u^2, it will be a straight line whose gradient is a and whose intercept on the uv-axis is b.

The table below gives the values from which the graph was constructed.

u	1	2	3	4	5
v	12·5	7·0	5·5	5·0	4·9
uv	12·5	14·0	16·5	20·0	24·5
u^2	1	4	9	16	25

The five points all lie on a straight line. By measuring the gradient and intercept on the uv-axis we find that $a = 1/2$ and $b = 12$.

[*The graph is left as an exercise for the reader. The resulting straight line should cut the uv-axis at* $uv = 12$, *and should pass through the point* (25, 24·5).]

Section 4 : Vectors

[*A vector represented both in magnitude and direction by the line joining the points P and Q in the sense from P to Q is denoted by $\vec{PQ}$ or by a single letter in bold type, e.g.* **F**. *The magnitude of the vector is denoted by PQ or F, and a unit vector parallel to* **F** *is denoted by* **F̂**. *The position vector of a point P relative to an origin O is the vector $\vec{OP}$ or* **r**. *The sum of vectors $\vec{PQ}$ and $\vec{QR}$ is $\vec{PR}$ ($\vec{PQ} + \vec{QR} = \vec{PR}$), and by definition $\vec{PQ} - \vec{QR} = \vec{PQ} + \vec{RQ}$. The vector $\lambda \vec{PQ}$ is parallel to the vector $\vec{PQ}$ and of magnitude λPQ.*

If **i**, **j** *and* **k** *are unit vectors in the x, y and z directions respectively,* **F** *may be expressed in the form*

$$\mathbf{F} = F_x \mathbf{i} + F_y \mathbf{j} + F_z \mathbf{k},$$

where F_x, F_y and F_z are the resolutes of **F** *in the x, y and z directions; thus* $\mathbf{r} = \mathbf{i} x + \mathbf{j} y + \mathbf{k} z$, *where P has cartesian coordinates (x, y, z). If F_x, F_y and F_z are each functions of a parameter, e.g. t, then*

$$\frac{d\mathbf{F}}{dt} = \frac{dF_x}{dt} \mathbf{i} + \frac{dF_y}{dt} \mathbf{j} + \frac{dF_z}{dt} \mathbf{k} \,.]$$

Question 5 :   (*G.C.E., Specimen Paper for 1977, A.*)

The vectors **a** and **b** are mutually perpendicular and $|\mathbf{a}| = 7$, $|\mathbf{b}| = 24$. Calculate $|\mathbf{a} + \mathbf{b}|$ and $|\mathbf{a} - \mathbf{b}|$.

Answer :

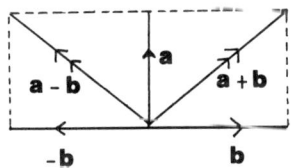

The construction of **a** + **b** and **a** − **b** is shown in the diagram. It is clear that

$$|\mathbf{a} + \mathbf{b}|^2 = |\mathbf{a} - \mathbf{b}|^2 = |\mathbf{a}|^2 + |\mathbf{b}|^2$$
$$= 7^2 + 24^2 = 625,$$

∴ $\quad |\mathbf{a} + \mathbf{b}| = |\mathbf{a} - \mathbf{b}| = 25.$

Question 6 : (*G.C.E., January 1975, A.*)

The points A, B have position vectors **a**, **b** respectively when referred to an origin O. Show that the vector equation of the line AB can be written in the form

$$\mathbf{r} = \mathbf{a} + t(\mathbf{b} - \mathbf{a}),$$

where **r** is the position vector of a point on the line and t is a parameter.

Answer :

For any point P on the line AB,

$\mathbf{r} = \overrightarrow{OP} = \overrightarrow{OA} + \overrightarrow{AP}.$

Now, $\overrightarrow{AP} = t\overrightarrow{AB}$ for some value of t, and $\overrightarrow{AB} = \mathbf{b} - \mathbf{a}$; thus

$$\overrightarrow{AP} = t(\mathbf{b} - \mathbf{a}).$$

Substitution into the first equation gives

$$\mathbf{r} = \mathbf{a} + t(\mathbf{b} - \mathbf{a}).$$

This is the vector equation of the line AB, for as t varies, P continues to lie on AB but at varying distances from A,

Q.E.D.

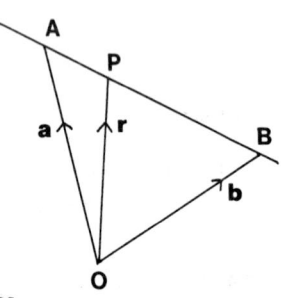

Section 5 : Variable Acceleration

[*For a particle of mass m moving in a straight line, and acted upon by a force $F(v)$ which is a function of the velocity v, the equation of motion is $a = m^{-1} F(v)$, where a is the acceleration. Writing $a = dv/dt$ and integrating expresses v as a function of t. Putting $v = ds/dt$, where s is the displacement, and integrating, gives s as a function of t. A relation between s and v may be obtained by writing $a = v \, dv/ds$ and integrating with respect to s.*]

Question 7 : (*G.C.E., Specimen Paper for 1977, A.*)

A particle moves in a straight line so that its velocity is directly proportional to the cube of the time for which it has been moving. After 2 seconds the particle has an acceleration of 3 m/s². Calculate the total distance moved in the first 3 seconds of its motion.

Answer :

If v is the velocity after time t,
$$v = kt^3 \text{ for some constant } k.$$

The acceleration is $dv/dt = 3kt^2$, and so
$$3 = 3k \cdot 2^2, \text{ whence } k = 1/4.$$

Since $v = \tfrac{1}{4}t^3$, the distance s moved in time t is given by
$$s = \int_0^t \tfrac{1}{4}t^3 \, dt = t^4/16,$$
and, after the first 3 seconds,
$$s = 3^4/16 = 5\tfrac{1}{16} \text{ m}.$$

Question 8 : (*G.C.E., Summer 1975, A.*)

A car of mass 1 000 kg moves along a horizontal road with acceleration proportional to the cube root of the time t seconds after starting from rest. When $t = 8$, the speed of the car is 8 m/s. Neglecting frictional resistances, calculate, in kW, the rate at which the engine driving the car is working when $t = 27$.

Answer :

Since the acceleration is proportional to $t^{1/3}$, the speed v satisfies the differential equation

$$dv/dt = kt^{1/3} \text{ for some constant } k. \qquad (1)$$

Thus
$$v = \tfrac{3}{4}kt^{4/3} + A, \qquad (2)$$

and since $v = 0$ when $t = 0$, $A = 0$. Also, since $v = 8$ when $t = 8$,

$$8 = \tfrac{3}{4}k \cdot 16, \text{ whence } k = 2/3. \qquad (3)$$

The pull of the engine is $1000\, dv/dt$, as the mass is 1 000 kg, and so the power P is given by

$$P = 1000\, v\, dv/dt \text{ W} = v\, dv/dt \text{ kW}$$
$$= \tfrac{3}{4}k^2 t^{5/3}, \qquad \text{from Equations (1) and (2).}$$

So at $t = 3$, using Equation (3),

$$P = \tfrac{3}{4}(\tfrac{2}{3})^2 3^5 = 81 \text{ kW}.$$

Question 9 : (*G.C.E., Specimen Paper for 1977, B.*)

(*i*) A particle P moves in a straight line Ox so that its acceleration at any instant is kv, where k is a positive constant and v is its velocity. The particle starts from O with velocity V at time $t = 0$. Show that

$$v = Ve^{kt}$$

and deduce an expression for OP at time t.

(*ii*) A particle P moves in a straight line Ox in the direction of x increasing so that $OP(=x)$ at time t is given by

$$x^2 = a^2 + V^2 t^2,$$

where a, V are constants.

Calculate the velocity and acceleration of P in terms of a, V and t.

Answer :

(*i*) The particle's motion satisfies the differential equation
$$dv/dt = kv.$$
Thus $\quad \int_V^v \dfrac{dv}{v} = k \int_0^t dt,$ since $v = V$ when $t = 0$.

Hence $\quad \log_e (v/V) = kt$,

and so $\quad v = Ve^{kt}$, Q.E.D.

If $OP = x$, we have $dx/dt = Ve^{kt}$,

and hence $\quad x = V \int_0^t e^{kt} \, dt$

since $x = 0$ when $t = 0$. It follows that
$$x = \dfrac{V}{k}(e^{kt} - 1).$$

(*ii*) On differentiating the given relation $x^2 = a^2 + V^2 t^2$ with respect to time, we obtain
$$2x(dx/dt) = 2V^2 t.$$
The velocity dx/dt is therefore given by
$$\dfrac{dx}{dt} = \dfrac{V^2 t}{x} = \dfrac{V^2 t}{(a^2 + V^2 t^2)^{1/2}}.$$

Differentiating the last equation with respect to t yields the acceleration d^2x/dt^2 in the form
$$\dfrac{d^2 x}{dt^2} = \dfrac{V^2}{(a^2 + V^2 t^2)^{1/2}} - \dfrac{V^4 t^2}{(a^2 + V^2 t^2)^{3/2}}$$
$$= \dfrac{a^2 V^2}{(a^2 + V^2 t^2)^{3/2}}.$$

Section 6 : Vectorial Kinematics

[*If the displacement of a point is represented vectorially by* $\mathbf{r} = \mathbf{i}x + \mathbf{j}y + \mathbf{k}z$, *then the velocity* $\mathbf{v}$ *is given by*

$$\mathbf{v} = d\mathbf{r}/dt = \mathbf{i}(dx/dt) + \mathbf{j}(dy/dt) + \mathbf{k}(dz/dt),$$

and the acceleration $\mathbf{a}$ *is given by*

$$\mathbf{a} = d\mathbf{v}/dt = d^2\mathbf{r}/dt^2 = \mathbf{i}(d^2x/dt^2) + \mathbf{j}(d^2y/dt^2) + \mathbf{k}(d^2z/dt^2).$$

If a particle of mass m is at $\mathbf{r}$, *then its momentum is* $\mathbf{P} = m\mathbf{v}$, *and the force* $\mathbf{F}$ *acting on it is*

$$\mathbf{F} = d\mathbf{P}/dt = m\, d\mathbf{v}/dt.$$

We sometimes write $\dot{x}$, $\dot{y}$ *for* dx/dt, dy/dt, *and* $\ddot{x}$, $\ddot{y}$ *for* d^2x/dt^2, d^2y/dt^2 *respectively.*]

Question 10 : (*G.C.E., Specimen Paper for 1977, A.*)

An aeroplane whose air speed is 600 km/h is to fly from A to B bearing 330° from A. The wind is blowing at 120 km/h from the West. Find by drawing or calculation the course the pilot must steer.

Answer :

In $\triangle ABC$, $\overrightarrow{AC}$ and $\overrightarrow{CB}$ represent respectively the wind's velocity and the aeroplane's velocity in the air. Thus $\overrightarrow{AB}$ represents the aeroplane's velocity relative to the ground. Since the bearing of B from A is 330°, $\angle BAD = 30°$, and so $\angle BAC = 120°$. Applying the sine rule to $\triangle ABC$, we have

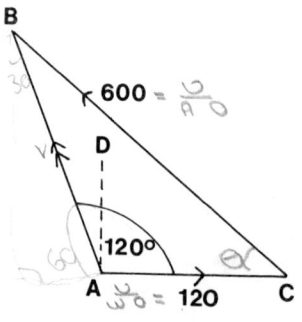

$$\frac{\sin \angle ABC}{120} = \frac{\sin 120°}{600},$$

whence $\sin \angle ABC = 0.1732$, i.e. $\angle ABC = 10°$. It follows that $\angle ACB = 180° - 10° - 120° = 50°$, and so the course that must be steered is $270° + 50° = 320°$.

Question 11 : (*G.C.E., January 1976, A.*)

The acceleration of a particle P moving in the plane of the coordinate axes Ox, Oy is $-n^2(x\mathbf{i} + y\mathbf{j})$, where n is constant and (x, y) are the coordinates of P. At time $t = 0$ the particle is projected with velocity $na\mathbf{j}$ from the point A, whose position vector is $2a\mathbf{i}$. Show that P describes the curve

$$\frac{x^2}{4a^2} + \frac{y^2}{a^2} = 1.$$

Answer :

The acceleration of P is $\ddot{x}\mathbf{i} + \ddot{y}\mathbf{j}$, and equating this to the given form yields

$$\ddot{x} = -n^2 x \quad \text{and} \quad \ddot{y} = -n^2 y$$

These equations each correspond to simple harmonic motion, and have the general solutions

$$x = A \sin(nt + \alpha), \quad y = B \sin(nt + \beta) \tag{1}$$
$$\dot{x} = nA \cos(nt + \alpha), \quad \dot{y} = nB \cos(nt + \beta), \tag{2}$$

where A or B and α or β are the amplitude and epoch of the motion respectively.

Now, at $t = 0$, $\mathbf{r} = 2a\mathbf{i}$ and $\dot{\mathbf{r}} = na\mathbf{j}$. Thus, at $t = 0$,
$x = 2a$, $y = 0$, $\dot{x} = 0$, $\dot{y} = na$.

Substituting these into Equations (1) and (2),

$$2a = A \sin \alpha, \quad 0 = nA \cos \alpha \tag{3}$$
$$0 = B \sin \beta, \quad na = nB \cos \beta. \tag{4}$$

From the second of Equations (3), $\alpha = \pi/2$, and so from the first $A = 2a$. From the first of Equations (4), $\beta = 0$, and from the second, $B = a$. Thus we have $x = 2a \cos nt$ and $y = a \sin nt$.

Hence $\qquad\qquad \dfrac{x^2}{4a^2} + \dfrac{y^2}{a^2} = 1, \qquad\qquad$ Q.E.D.

Question 12: (*G.C.E., January 1975, A.*)

A particle P of mass m moves so that its position vector $\overrightarrow{OP} = \mathbf{r}$ at time t is given by

$$\mathbf{r} = \mathbf{a} \cos \omega t + \mathbf{b} \sin \omega t,$$

where $\mathbf{a}$, $\mathbf{b}$ are constant non-parallel vectors, $|\mathbf{a}| > |\mathbf{b}|$ and $\omega \, (\neq 0)$ is a constant. Show that P lies in a fixed plane which passes through O. Show also that the resultant force $\mathbf{F}$ acting on P is directed towards O.

Answer:

Consider the given vectors $\mathbf{a}$, $\mathbf{b}$, and the vector $\overrightarrow{OP} = \lambda \mathbf{a} + \mu \mathbf{b}$ for any pair of scalars λ and μ. It is clear that for all λ and μ $\overrightarrow{OP}$ lies in the plane defined by the vectors $\mathbf{a}$ and $\mathbf{b}$.

Hence, if P has position vector $\mathbf{r}$ given by

$$\mathbf{r} = \mathbf{a} \cos \omega t + \mathbf{b} \sin \omega t,$$

then P lies on a fixed plane passing through O, $\qquad$ Q.E.D.

The force $\mathbf{F}$ acting on P is given by

$$\mathbf{F} = md^2\mathbf{r}/dt^2 = -\omega^2 \mathbf{a} \cos \omega t - \omega^2 \mathbf{b} \sin \omega t$$
$$= -\omega^2 \mathbf{r}.$$

Thus $\mathbf{F}$ is parallel to $\overrightarrow{OP}$ and acts in the opposite sense. It is therefore directed towards O, $\qquad$ Q.E.D.

Question 13: (*G.C.E., Summer 1975, B.*)

Two particles A and B move with constant velocity vectors $(4\mathbf{i} + \mathbf{j} - 2\mathbf{k})$ and $(6\mathbf{j} + 3\mathbf{k})$ respectively, the unit of speed being the metre per second. At time $t = 0$, A is at the point with position vector $(-\mathbf{i} + 20\mathbf{j} + 21\mathbf{k})$ and B is at the point with position vector $(\mathbf{i} + 3\mathbf{k})$, the unit of distance being the metre.

Find the value of t for which the distance between A and B is least and find also this least distance.

Answer :

If A is at Q when $t = 0$, then

$$\vec{OA} = \vec{OQ} + \vec{QA},$$

where $\vec{OQ} = -\mathbf{i} + 20\mathbf{j} + 21\mathbf{k}$

and $\vec{QA} = (4\mathbf{i} + \mathbf{j} - 2\mathbf{k})t$.

Thus $\vec{OA} = \mathbf{i}(-1 + 4t) + \mathbf{j}(20 + t) + \mathbf{k}(21 - 2t)$.

Similarly $\vec{OB} = \mathbf{i} + 6t\mathbf{j} + 3(1 + t)\mathbf{k}$.

Now, if $\vec{AB} = \mathbf{r}$,

$$\mathbf{r} = \vec{OB} - \vec{OA} = \mathbf{i}(2 - 4t) + \mathbf{j}(-20 + 5t) + \mathbf{k}(-18 + 5t).$$

Hence $r = [(2 - 4t)^2 + (-20 + 5t)^2 + (-18 + 5t)^2]^{1/2}$
$= [2(33t^2 - 198t + 364)]^{1/2}$. (1)

So r will be minimized when $33t^2 - 198t + 364$ is minimized; differentiating with respect to t and equating the result to zero gives

$$66t - 198 = 0, \text{ whence } t = 3 \text{ seconds}.$$

(Since the above quadratic form has a positive coefficient of t^2, this stationary point is indeed a minimum.)

On substituting $t = 3$ into Equation (1), we obtain $r = \sqrt{134}$ metres.

Question 14 : (*G.C.E., Summer 1974, B.*)

Two particles A and B have position vectors $(2 \sin \omega t)\mathbf{i} + (2 \cos \omega t)\mathbf{j}$, where $\omega > 0$, and $2t\mathbf{i} + t^2\mathbf{j}$ respectively at time t. Find the cartesian equations of the paths followed by A and B.

Find (a) the magnitude of the velocity of A relative to B when $t = 0$,

(b) the magnitude of the acceleration of A relative to B when $t = \pi/(2\omega)$.

Find also the values of t for which

(c) the accelerations of A and B are parallel and in the same sense,

(d) the accelerations of A and B are parallel and in opposite senses.

Answer:

If (x, y) are the coordinates of A, it follows from the given equations that

$$x = 2 \sin \omega t \quad \text{and} \quad y = 2 \cos \omega t.$$

Thus $x^2 + y^2 = 4$, and this is the required cartesian equation of A's path.

Similarly for B,

$$x = 2t \quad \text{and} \quad y = t^2;$$

therefore $x^2 = 4t^2 = 4y$, and so $x^2 = 4y$ is the required equation for B.

(a) Differentiating with respect to time, we find the velocities $\mathbf{v}_A$ and $\mathbf{v}_B$ at time t to be

$$\mathbf{v}_A(t) = 2\omega \cos \omega t \, \mathbf{i} - 2\omega \sin \omega t \, \mathbf{j}$$

and

$$\mathbf{v}_B(t) = 2\mathbf{i} + 2t \, \mathbf{j}.$$

The velocity of A relative to B at $t = 0$ is

$$\mathbf{V} = \mathbf{v}_A(0) - \mathbf{v}_B(0)$$
$$= 2(\omega - 1)\mathbf{i},$$

and so

$$V = 2(\omega - 1).$$

(b) A further differentiation gives the accelerations $\mathbf{a}_A(t)$ and $\mathbf{a}_B(t)$ as

$$\mathbf{a}_A(t) = -2\omega^2 \sin \omega t \, \mathbf{i} - 2\omega^2 \cos \omega t \, \mathbf{j}$$

and $\quad \mathbf{a}_B(t) = 2\mathbf{j}$.

The acceleration of A relative to B at $t = \pi/2\omega$ is

$$\mathbf{f} = \mathbf{a}_A(2\pi/\omega) - \mathbf{a}_B(2\pi/\omega)$$
$$= -2\omega^2 \mathbf{i} - 2\mathbf{j}.$$

Thus $\quad f = 2(\omega^4 + 1)^{1/2}$.

If the accelerations of A and B are parallel at time t, the coefficient of $\mathbf{i}$ in $\mathbf{a}_A(t)$ must be zero, and so $\sin \omega t = 0$, giving

$$\omega t = n\pi \text{ for any integer } n.$$

If n is odd, $\cos \omega t = -1$ and therefore $\mathbf{a}_A = +2\omega^2 \mathbf{j}$, corresponding to $\mathbf{a}_A$ being in the same sense as $\mathbf{a}_B$. If n is even, $\cos \omega t = +1$ and $\mathbf{a}_A = -2\omega^2 \mathbf{j}$, corresponding to $\mathbf{a}_A$ being in the opposite sense to $\mathbf{a}_B$. Hence

(c) $t = n\pi/\omega$ with n odd;

(d) $t = n\pi/\omega$ with n even.

Section 7 : Laws of Motion

[*If a particle moving in a straight line with constant acceleration f and initial velocity u reaches a velocity v in time t, after moving a distance s, then $v = u + ft$, $v^2 = u^2 + 2fs$, and $s = ut + \frac{1}{2}ft^2$.*

The motion of a system of bodies acted upon by constant forces may be investigated by applying to each particle the law: *force = mass × acceleration. In the absence of dissipative forces, such as friction, application of the law of conservation of energy can sometimes give the required result more quickly.*]

Question 15 : (*G.C.E., Summer 1976, A.*)

A man standing on a platform throws a ball vertically upwards with velocity v. Immediately after the ball leaves his hand, the man and the platform descend vertically with constant velocity kv. Show that the time that elapses before the ball returns to his hand is $2(k + 1)v/g$.

Answer :

If a ball is thrown vertically upwards with velocity V, then the time for it to return to a stationary man is $2V/g$. If the ball is thrown upwards with velocity v, and the man descends with constant velocity kv, then the initial vertical velocity of the ball *relative* to the man is $kv + v = (k + 1)v$. The total time of flight is therefore given by taking $V = (k + 1)v$ in the above result $2V/g$. This gives a total time of $2(k + 1)v/g$,   Q.E.D.

Question 16 : (*G.C.E., Specimen Paper for 1977, B.*)

Two particles A and B of mass M and m respectively, where $M > m$, are connected by a light inextensible string passing over a smooth fixed pulley. They are released from rest with the hanging parts of the string vertical. Find the force exerted by the string on the pulley.

LAWS OF MOTION

After time t_1 the string breaks just as the particles are passing each other. The particle A strikes the ground after a further time t_2. Prove that the particle B is then at a height

$$2(M - m)gt_1t_2/(M + m)$$

above A.

Answer:

Let T be the tension in the string and a the acceleration of the masses as shown in the diagram. Then applying $force = mass \times acceleration$ to A and B in turn gives

$$Mg - T = Ma$$

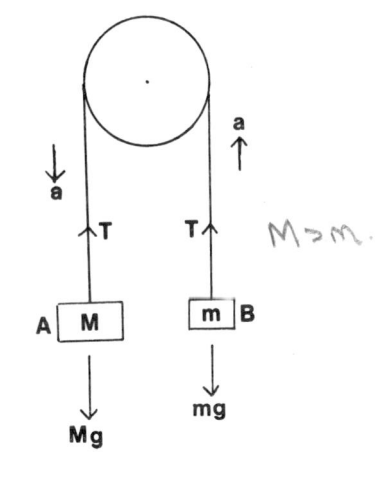

and $\qquad T - mg = ma$.

Thus $\qquad (M - m)g = (M + m)a$,

and so $\qquad a = (M - m)g/(M + m)$.

Substituting for a into either of the above equations yields

$$T = 2Mmg/(M + m);$$

the force exerted by the string on the pulley is

$$2T = 4Mmg/(M + m).$$

After the string breaks, the particles both move vertically with acceleration g, so their relative acceleration is zero. If V is the speed of each particle when the string breaks, their relative speed is $2V$, and this will remain constant since their relative acceleration is zero. The distance h between them after time t_2 is thus $2Vt_2$, where $V = at_1$. Substituting for a from the above result gives finally

$$h = \frac{2(M - m)gt_1t_2}{(M + m)}, \qquad Q.E.D.$$

Question 17 : (*G.C.E., Summer 1974, B.*)

A light inextensible string passes over a fixed light smooth pulley and carries scale pans A and B, each of mass 4 kg, at its ends. An inelastic ball of mass 2 kg is placed in the scale pan A. The system is released from rest with the string taut and the straight parts of the string vertical. Find

(*a*) the tension, in newtons, in the string,
(*b*) the acceleration, in m/s^2, of the ball,
(*c*) the force, in newtons, exerted by the ball on the scale pan A,
(*d*) the force, in newtons, exerted by the string on the pulley.

Two seconds after the commencement of the motion scale pan A strikes an inelastic horizontal plane. Calculate

(*e*) the further time which elapses before A is once again jerked into motion,
(*f*) the impulse in the string when A is jerked into motion. (Take g as 9·8 m/s^2.)

Answer :

(*a*) and (*b*) Let T be the tension in the string and f be the acceleration of the ball. Applying *force = mass × acceleration* to each side in turn:

$$6g - T = 6f \qquad (1)$$
and $\qquad T - 4g = 4f. \qquad (2)$
By addition, $\qquad 2g = 10f,$
whence $f = g/5 = 1\cdot96\,\mathrm{m\,s^{-2}}$.

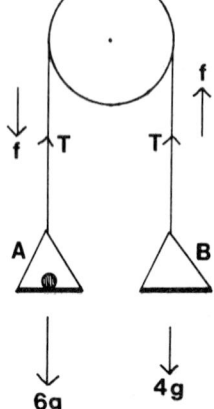

Substitution in Equation (2) yields
$$T = 4g + 4g/5$$
$$= 24g/5 = 47\cdot04\,\mathrm{N}.$$

LAWS OF MOTION

(c) If R is the reaction on the ball by the scale,
$$2g - R = 2f,$$
and so
$$R = 2(g - \tfrac{1}{5}g) = 8g/5$$
$$= 15 \cdot 68 \text{ N}.$$

(d) The force exerted on the pulley by the string is
$$2T = 94 \cdot 08 \text{ N}.$$

(e) Since the acceleration of the string is $g/5$, the speed of B after 2 s is $2g/5$. When A strikes the plate it stops, and B then moves freely under gravity with initial vertical velocity $V = 2g/5$. The time of flight is
$$t = 2V/g = (4/5) \text{ s},$$
and this is the time before A is jerked into motion.

(f) When B falls back to its 'point of projection' it will have velocity V downwards, and hence momentum $4V$. This momentum will then be shared with the mass of 6 kg at A, and if v is the resulting speed, conservation of momentum gives
$$4V = (4+6)v, \quad \text{whence} \quad v = (2/5)V.$$

The impulse in the string will be the momentum J transmitted to the mass at A. We have
$$J = 6v = (12/5)V$$
$$= 24g/25, \text{ using the value obtained above for } V.$$
Hence $\quad J = 9 \cdot 408 \text{ kg m s}^{-1}.$

Question 18 : (*G.C.E., Summer 1976, B.*)

Two particles, of masses 1·2 kg and 1·3 kg, are connected by a light inextensible string which passes over a fixed light smooth pulley. The system is released from rest with the string taut and the straight parts of the string vertical. Calculate the acceleration of each particle and the tension in the string.

When 2 seconds have elapsed after the system starts from rest, the lighter particle picks up a mass of 1·0 kg which was at rest before being picked up. Calculate the further time that elapses before the system comes instantaneously to rest and the total distance that the lighter particle has moved. (Take g as 10 m/s².)

Answer :

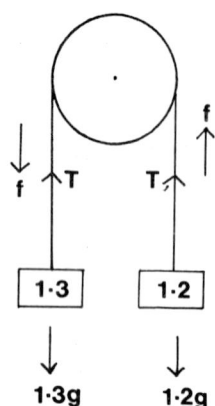

Applying *force = mass × acceleration* to each mass in turn:

$$1·3g - T = 1·3f \quad (1)$$

and $\quad T - 1·2g = 1·2f. \quad (2)$

Adding these equations,

$$0·1g = 2·5f,$$

and so $\quad f = (2/5) \text{ ms}^{-2}, \quad (3)$

since $\quad g = 10 \text{ ms}^{-2}.$

From Equation (2),

$$T = 1·2(g + f) = 1·2 \times 10·4$$
$$= 12·48 \text{ N}.$$

It follows from Equation (1) that after 2 seconds the speed of the masses is $(4/5)$ ms^{-1}. Momentum is conserved, so the new velocity v after the stationary mass of 1·0 kg has been picked up will be given by

$$(1·3 + 1·2)(4/5) + 1 \times 0 = (1·3 + 1·2 + 1)v \, ;$$

hence $\qquad\qquad\qquad v = (4/7) \text{ ms}^{-1}.$

LAWS OF MOTION

For the subsequent motion the masses are 1·3 kg and 2·2 kg. If T_1 and f_1 are the new tension and acceleration, then

$$2\cdot 2g - T_1 = 2\cdot 2 f_1$$

and

$$T_1 - 1\cdot 3g = 1\cdot 3 f_1 \ .$$

Adding these equations,

$$0\cdot 9 g = 3\cdot 5 f_1 \ , \text{ whence } f_1 = (18/7) \text{ m s}^{-2} \ .$$

Since the initial velocity is $(4/7)$ m s^{-1}, the time t to come instantaneously to rest is given by

$$(4/7) - (18/7)t = 0, \text{ so that } t = (2/9) \text{ s} \ .$$

The initial distance moved before picking up the 1·0 kg mass is

$$S_1 = \tfrac{1}{2} f \cdot 2^2$$
$$= (4/5) \text{ m from Equation (3)} \ .$$

The distance moved after picking up the 1·0 kg mass is

$$s_2 = \tfrac{1}{2} f_1 \cdot (2/9)^2$$
$$= (4/63) \text{ m} \ .$$

Thus the total distance moved is

$$s_1 + s_2 = (272/315) \text{ m} \ .$$

Section 8 : Collisions and Momentum

[*If only internal forces act upon a system, its momentum is constant. If the momentum is changed by the action of external forces, the impulse of these forces is equal to the change in momentum they produce.*

Thus for the collision of two spheres moving along the line of their centres, the total momentum is unchanged. In addition, the restitution law states that the relative velocity after impact along the common normal equals $-e$ times the relative velocity before impact, where e is the coefficient of restitution. Note that $0 \leqslant e \leqslant 1$. Kinetic energy is always lost during a collision unless $e = 1$, when it is conserved.]

Question 19 : (*G.C.E., Specimen Paper for 1977, A.*)

A sphere A of mass $2m$ moving with velocity $2V$ along a smooth horizontal table overtakes and collides with a second sphere B, of the same radius and mass m and moving with velocity V along the same horizontal line. As a result of the impact, the velocity of B is doubled. Calculate the coefficient of restitution between the spheres.

Answer :

Since the velocity of B is doubled, after impact its velocity will be $2V$. Let the velocity of A after impact be W as shown. Then, because momentum is conserved,

$$2m \cdot 2V + mV = 2mW + m \cdot 2V \ ;$$

that is, $\qquad\qquad 3V = 2W \ .$ \hfill (1)

Applying the restitution law,

$2V - W = e(2V - V),$ \quad i.e. \quad $2V - W = eV \ .$ \hfill (2)

Substituting for W from Equation (1) into Equation (2) gives

$$2V - (3V/2) = eV, \ \text{ whence } \ e = 1/2 \ .$$

COLLISIONS AND MOMENTUM

Question 20: (*G.C.E., Summer 1974, A.*)

A bullet of mass 0·02 kg is fired horizontally with a velocity of 505 m/s directly into a wooden block of mass 2 kg, which is free to move on a smooth horizontal table, and becomes embedded in the block. Find, in joules, the loss of kinetic energy. (You may assume that the block does not rotate.)

Answer:

Let v be the speed of the block with embedded bullet. The conservation of momentum requires that

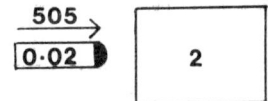

$$0{\cdot}02 \times 505 = 2{\cdot}02 \times v,$$

i.e. that $v = 5 \text{ ms}^{-1}$.

Initial kinetic energy $= 0{\cdot}01 \times 505^2 \text{ J}$, $\tfrac{1}{2}mv^2$

and final kinetic energy $= 1{\cdot}01 \times 5^2 \text{ J}$,

∴ loss of kinetic energy $= 0{\cdot}01 \times 505^2 - 1{\cdot}01 \times 5^2 \text{ J}$
$$= 2525 \text{ J}.$$

Question 21: (*G.C.E., Summer 1975, A.*)

A shell of mass m is fired horizontally by a gun of mass km which is free to recoil on horizontal ground. The total kinetic energy imparted to the shell and the gun together is E. Calculate, in terms of k, m and E, the speed of the shell at the instant when it leaves the gun.

Answer:

Let the shell be fired with speed u and let v be the recoil speed of the gun. Then the principle of conservation of momentum gives

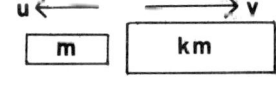

$$mu = kmv, \qquad (1)$$

We have $\qquad E = \tfrac{1}{2}mu^2 + \tfrac{1}{2}kmv^2$,

and substituting in this for v from Equation (1) yields

$$E = \tfrac{1}{2}mu^2 + \tfrac{1}{2}km(u/k)^2$$

$$= \tfrac{1}{2}mu^2\left(1 + \tfrac{1}{k}\right).$$

Thus $\qquad u = \left(\dfrac{2kE}{m(1+k)}\right)^{1/2}.$

Question 22 : (*G.C.E., Summer 1976, B.*)

Two spheres, A and B, of equal radii but of masses m and $3m$ respectively, lie at rest on a smooth horizontal floor. The spheres lie in a line perpendicular to a vertical wall, with B nearer to the wall than A. Sphere A is given a velocity u in the direction AB and strikes sphere B. Show that, if the coefficient of restitution between A and B is $\tfrac{1}{4}$, then B moves off with velocity $5u/16$.

Sphere B goes on to strike the wall and, after rebounding from the wall, is brought to rest by its second impact with A. Show that the coefficient of restitution between B and the wall is $1/11$.

Answer :

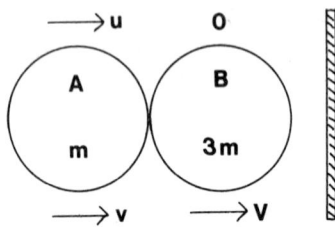

If the velocities after the first impact of A and B are respectively v and V, it follows from the conservation of momentum that

$$mu = mv + 3mV, \text{ and so } u = v + 3V. \qquad (1)$$

Application of the restitution law gives
$$\tfrac{1}{4}u = V - v, \qquad (2)$$
since the coefficient of restitution for this impact is $\tfrac{1}{4}$. Solving Equations (1) and (2) for v and V gives
$$v = u/16 \quad \text{and} \quad V = 5u/16, \qquad Q.E.D.$$

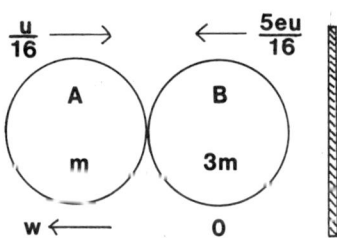

Let the coefficient of restitution between B and the wall be e. Then, after collision with the wall, B will be moving in the opposite direction with speed $eV = 5eu/16$. A subsequent collision with A brings B to rest. Let the speed of A after this impact be w; applying the principle of conservation of momentum yields
$$3m \cdot \frac{5eu}{16} - m \cdot \frac{u}{16} = mw,$$
and so $\qquad (15eu/16) - (u/16) = w.\qquad (3)$

The restitution law gives
$$w = \frac{1}{4}\left(\frac{u}{16} + \frac{5eu}{16}\right),$$
whence $\qquad 64w = u(1 + 5e). \qquad (4)$

From Equations (3) and (4),
$$u(1 + 5e) = 64w = 60eu - 4u,$$
whence $\qquad 55e = 5, \text{ i.e. } e = 1/11, \qquad Q.E.D.$

Question 23 : (*G.C.E., Summer 1975, B.*)

Three small smooth spheres *A*, *B*, *C* of equal radii and of masses m, $2m$, $4m$ respectively, lie at rest and separated from one another on a smooth horizontal table in the order *A*, *B*, *C* with their centres in a straight line. The coefficient of restitution between any two spheres is *e*. Sphere *A* is projected with speed *V* directly towards sphere *B*. Find the velocity of each of the spheres just after *C* is set in motion.

If $e \geqslant \frac{1}{2}$, show that only two collisions take place.

Answer :

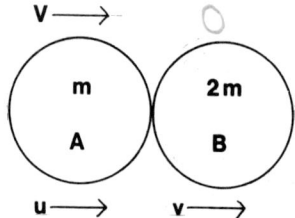

For the first collision it follows from the conservation of momentum that

$$mV = mu + 2mv, \quad \text{and so} \quad V = u + 2v. \quad (1)$$

The restitution law gives $\qquad eV = v - u.\qquad (2)$

Solving Equations (1) and (2) for *u* and *v*, we find

$$u = \left(\frac{1-2e}{3}\right)V \quad \text{and} \quad v = \left(\frac{1+e}{3}\right)V. \quad (3)$$

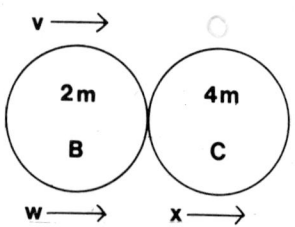

For the second collision, C is initially stationary, and *B* approaches it with speed *v*. Since the ratio of the mass of *C* to that of B is the same as the ratio of the mass of B to that of A, it follows that the velocities w and x after the second collision will bear the same ratio to v as did u and v to V. Thus

$$w = \left(\frac{1-2e}{3}\right)v = \frac{(1-2e)(1+e)}{9}V \qquad (4)$$

and $$x = \left(\frac{1+e}{3}\right)v = \frac{(1+e)^2}{9}V. \qquad (5)$$

So, after C is set in motion the velocities of A, B and C will be u, w and x respectively, given in Equations (3), (4) and (5).

If $e \geqslant \frac{1}{2}$, then $1 - 2e \leqslant 0$. It follows from the above results that $u \leqslant 0$, $w \leqslant 0$ and $x > 0$. Thus neither A nor B can collide with C. Also,

$$\frac{w}{u} = \frac{1+e}{3} < 1 \text{ (since } e \leqslant 1).$$

Therefore $|w| < |u|$, and so A cannot collide again with B. This completes the proof that only two collisions occur,

Q.E.D.

Section 9 : Energy and Power

[*The work done by a force is equal to the product of the force and the distance moved in the direction of the force. The work performed on a system is equal to the increase in the energy of the system.*

Power is the rate of doing work, and is equal to the product of the force and the component of velocity in the direction of the force.

The unit of work is the joule and the unit of power is the watt, equal to one joule per second.]

MATHEMATICS (APPLIED)

Question 24 : (G.C.E., Summer 1975, A.)

A pump raises 100 kg of water per second from a depth of 30 m. The water is delivered at a speed of 30 m/s. Find, in joules, (a) the potential energy, (b) the kinetic energy gained by the water delivered each second. Neglecting frictional losses calculate, in kW, the rate at which the pump is working. (Take g as 9·8 m/s².)

Answer :

(a) The weight of water raised per second is $100\,g$ N, and since this is raised through 30 m, the potential energy it acquires is

$$100g \times 30 = 29400 \text{ J}.$$

(b) Since the velocity of the water is 30 m s^{-1}, the kinetic energy given to it each second is

$$\tfrac{1}{2} \cdot 100 \cdot 30^2 = 45000 \text{ J}.$$

The total work done by the pump per second is thus

$$29400 + 45000 = 74400 \text{ J},$$

and so its power is 74400 W = 74·4 kW.

Question 25 : (G.C.E., Specimen Paper for 1977, B.)

A car of mass 1 000 kg has a maximum speed of 15 m/s up a slope inclined at an angle θ to the horizontal where $\sin \theta = 0\cdot 2$. There is a constant frictional resistance equal to one tenth of the weight of the car. Find the maximum speed of the car on a level road assuming that the engine works at the same rate.

If the car descends the same slope with its engine working at half this rate, find the acceleration of the car at the moment when its speed is 30 m/s. (Take g as 9·8 m/s².)

ENERGY AND POWER

Answer:

When the car is moving up the slope, the total force down the slope is the sum of the resolute of the weight, $Mg \sin \theta$, and the frictional force, $0.1\,Mg$. Since power = force × velocity, this gives

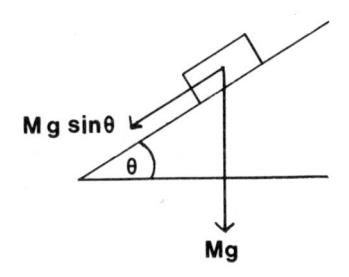

$$(Mg \sin \theta + 0.1\,Mg)15 = P \qquad (1)$$

where P is the power of the engine. Let v be the maximum speed on a level road; then

$$0.1\,Mgv = P. \qquad (2)$$

It follows from Equations (1) and (2) that

$$\frac{v}{15} = \frac{0.2 + 0.1}{0.1} = 3$$

(since $\sin \theta = 0.2$), and so $v = 45\,\text{ms}^{-1}$.

When the car descends the slope with engine power $P/2$, the net force down the slope will be

$$P/(2 \times 30) + Mg \sin \theta - 0.1\,Mg,$$

and equating this to mass × acceleration gives

$$(P/60) + 0.1\,Mg = Ma. \qquad (3)$$

From Equation (1), $P = 4.5\,Mg$, and so Equation (3) yields

$$(7/40)Mg = Ma,$$

whence
$$a = \frac{7 \times 9.8}{40} = 1.715\,\text{ms}^{-2}.$$

Question 26 : (G.C.E., Summer 1976, B.)

The maximum rate of working of the engine of a car is S kW. Against a constant resistance, the car can attain a maximum speed of u m/s on level ground and a maximum speed of $\frac{1}{2}u$ m/s directly up a slope of inclination α where $\sin \alpha = \frac{1}{16}$. Calculate the maximum speed of the car up a slope of inclination β, where $\sin \beta = \frac{1}{8}$, assuming that the resistance remains unchanged.

Given that $S = 15$ and that $u = 20$, find the maximum acceleration that can be attained when the car is towing a trailer of mass 300 kg at 10 m/s on level ground. It may be assumed that the resistance to the motion of the car remains unchanged and that the resistance to the motion of the trailer can be neglected. (Take g as 10 m/s².)

Answer :

If the constant resistance is R newtons, then since the engine power is $1000\, S$ W,

$$1000\, S = Ru\,. \tag{1}$$

When the car is ascending a slope of inclination α, the total downward force is $R + mg \sin \alpha$, and so

$$1000\, S = (R + mg \sin \alpha)(u/2)\,. \tag{2}$$

If v is the maximum speed up a slope of inclination β, it follows similarly that

$$1000\, S = (R + mg \sin \beta)v\,. \tag{3}$$

Equating the right-hand sides of Equations (1) and (2) gives

$$R = mg \sin \alpha = mg/16\,. \tag{4}$$

Equating the right-hand sides of Equations (1) and (3) yields

$$Ru = (R + mg \sin \beta)v\,,$$

whence
$$v = \frac{1/16}{(1/16) + (1/8)}\, u$$

since $\sin \beta = 1/8$ and $R = mg/16$.

Thus $\qquad v = u/3$.

ENERGY AND POWER

If $S = 15$ and $u = 20$, it follows from Equation (1) that
$$R = \frac{15000}{20} = 750,$$
and from Equation (4) that
$$m = \frac{16R}{g} = 1200.$$

The total mass, including the trailer, is therefore
$$M = 1200 + 300 = 1500 \text{ kg},$$
and the force exerted by the engine is
$$P = 15000/10 = 1500 \text{ N}.$$

Hence the net tractive force is
$$F = P - R = 750 \text{ N};$$
and if a is the maximum acceleration, we therefore have $F = Ma$, whence
$$750 = 1500\,a, \quad \text{i.e.} \quad a = (1/2) \text{ ms}^{-2}.$$

Question 27: (*G.C.E., Summer 1974, B.*)

A train is travelling along a straight level track at a speed of 36 km/h. If the power exerted by the engine is 500 kW find, in newtons, the tractive force exerted by the engine. If the train has total mass 6×10^5 kg and has an acceleration of 0·05 m/s², calculate the total resistance to the motion.

The mass of the engine alone is 10^5 kg. Assuming that the resistance to the motion of any part of the train is proportional to its mass, find the tension in the coupling between the engine and the rest of the train.

Answer :

$$36 \text{ km h}^{-1} = \frac{36 \times 1000}{3600} \text{ ms}^{-1} = 10 \text{ ms}^{-1}.$$

The engine power is 500×1000 W, and so the tractive force is

$$P = \frac{500 \times 1000}{10} = 5 \times 10^4 \text{ N}.$$

Since the mass is 6×10^5 kg and the acceleration is 0.05 ms^{-2}, the force required to produce the acceleration is

$$T = 6 \times 10^5 \times 0.05 = 3 \times 10^4 \text{ N}.$$

The resistance is $P - T$, and is therefore 2×10^4 N.

Since the engine mass is 10^5 kg, the mass of the rest of the train is 5×10^5 kg, i.e. 5/6 of the total mass. Therefore the resistance to motion of the rest of the train will be 5/6 of the total resistance, and the accelerating force on it will be 5/6 of the total accelerating force (since accelerating force is proportional to mass). Thus the tractive force on the rest of the train is $T_1 = (5/6)P$, and this is the tension in the coupling between the engine and the rest of the train. Hence

$$T_1 = (5/6) \times 5 \times 10^4 = 41666\tfrac{2}{3} \text{ N}.$$

Section 10 : Motion of a Particle in a Plane

[*If points P and Q have velocities* **u** *and* **v** *respectively, the velocity of P relative to Q is* **u** − **v**.

If a particle is projected with initial velocity u at an angle α to the horizontal, the horizontal distance travelled after time t is $x = ut \cos \alpha$, and the vertical displacement is $y = ut \sin \alpha - \frac{1}{2}gt^2$. Elimination of t gives the cartesian equation of the trajectory as $y = x \tan \alpha - (gx^2 \sec^2\alpha / 2u^2)$, which is a parabola. The maximum height from and range along the horizontal plane are $u^2 \sin^2 \alpha / 2g$ and $u^2 \sin 2\alpha / g$ respectively.

A particle of mass m moving at speed v in a circle of radius r has an acceleration towards the centre of $v^2/r = r\omega^2$, where ω is the angular velocity $(=v/r)$. The force towards the centre is therefore $mr\omega^2$.]

Question 28 : (*G.C.E., Specimen Paper for 1977, A.*)

A particle moving inside a smooth spherical bowl of radius $4a$ is describing a horizontal circle at a distance $2a$ below the centre of the bowl. Prove that its speed is $\sqrt{(6ga)}$.

Answer :

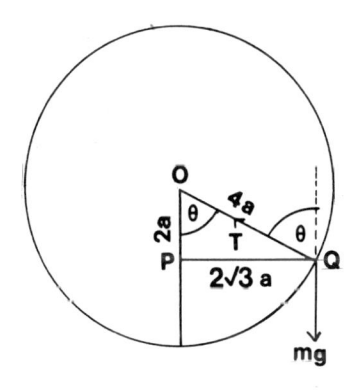

Cos $\theta = 2a/4a = 1/2$, and thus $\theta = 60°$. Resolving vertically the forces acting on the particle,

$$mg = T \cos 60° = \tfrac{1}{2}T.$$

Resolving horizontally,

$$\frac{mv^2}{2a\sqrt{3}} = T\cos 30° = \frac{T\sqrt{3}}{2}$$

$$= mg\sqrt{3} \text{ (from previous equation)}.$$

Thus $\quad v^2 = 6ag,$ whence $v = \sqrt{(6ag)},\qquad Q.E.D.$

Question 29 : (*G.C.E., Summer 1974, A.*)

A particle of mass 2 kg is attached to the mid-point B of an inextensible string ABC, and another particle of mass 3 kg is attached to the end C. The particles revolve uniformly round the end A, which is fixed, upon a smooth horizontal table, so that ABC is always a straight line. Show that the tensions in AB, BC are in the ratio $4:3$.

Answer :

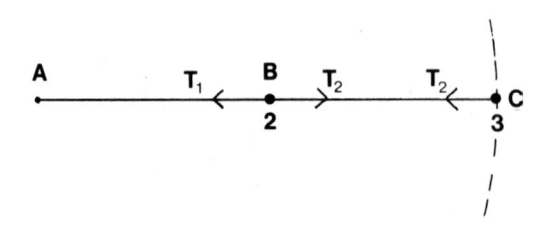

Let T_1 be the tension in the portion of string AB, and T_2 that in the portion BC. Then the force on C towards the centre is T_2, and that on B is $T_1 - T_2$. If $AB = a$, applying *force = mass × acceleration* to B and C respectively gives

$$T_1 - T_2 = 2a\omega^2$$
and $\qquad\qquad T_2 = 3 \times 2a\omega^2,$

where ω is the angular velocity of AC.

Thus $\qquad\qquad T_2 = 3(T_1 - T_2),$
whence $\qquad\qquad T_1/T_2 = 4/3,\qquad\qquad Q.E.D.$

MOTION OF A PARTICLE IN A PLANE

Question 30 : (*G.C.E., Summer 1976, B.*)

A yacht A is sailing due east at 9 knots and a second yacht B is sailing on a bearing of 030° at 6 knots. At a certain instant a third yacht C appears to an observer on A to be sailing due south and appears to an observer on B to be sailing on a bearing of 150°. Find the speed of the yacht C and the bearing on which it is sailing. (1 knot = 1 nautical mile per hour.)

Answer :

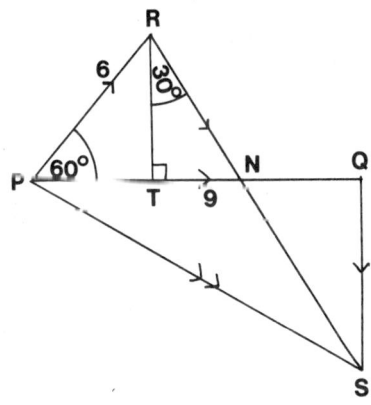

Let $\overrightarrow{PQ}$, $\overrightarrow{PR}$ and $\overrightarrow{PS}$ represent respectively the velocities of A, B and C. Then since C appears to A to be sailing due south,

$$\overrightarrow{PQ} + \overrightarrow{QS} = \overrightarrow{PS},$$

where $\overrightarrow{QS}$ is due south. Also, since C appears to B to be sailing on a bearing of 150°,

$$\overrightarrow{PR} + \overrightarrow{RS} = \overrightarrow{PS},$$

where $\angle SRT = 30°$. We require to find PS and $\angle QPS$.

Now, since $\angle PRT = 30°$, $\triangle PRN$ is equilateral, so that $PN = 6$ and therefore $QN = 9 - 6 = 3$. Also, $\angle NSQ = 30°$ (since $RT \parallel QS$), and so $QS = QN/\tan 30° = 3\sqrt{3}$. It follows from Pythagoras' Theorem that

$$PS = (9^2 + 9 \cdot 3)^{1/2} = 108^{1/2} = 6\sqrt{3}.$$

Further, $\tan \angle QPS = 3\sqrt{3}/9 = 1/\sqrt{3}$, whence $\angle QPS = 30°$. The speed of C is thus $6\sqrt{3}$ knots and its course is $90° + 30° = 120°$.

Question 31 : (*G.C.E., Summer 1976, B.*)

A golf ball, driven from a point P with an initial speed of 50 m/s, first strikes the ground at a point Q on the same horizontal level as P and 200 m from P. Neglecting air resistance, find, correct to the nearest degree, each of the two possible angles of projection and, correct to the nearest tenth of a second, the difference in the corresponding times of flight.

Show that, in its lower trajectory, the ball could not clear a tree 30 m high anywhere in its line of flight. (Take g as 10 m/s².)

Answer :

If a ball is projected with speed V at an angle α to the horizontal, the time of flight $T = (2V \sin \alpha)/g$, and the range $R = [(2V \sin \alpha)/g] V \cos \alpha = (V^2 \sin 2\alpha)/g$. Thus in this case,

$$200 = (50^2 \sin 2\alpha)/10, \quad \text{whence} \quad \sin 2\alpha = 4/5 .$$

This gives $2\alpha = 53° 08'$ or $180° - 53° 08'$,

i.e. $\alpha = 26° 34'$ or $63° 26'$.

To the nearest degree, the two possible angles of projection are therefore 27° and 63°.

Since $T = (2V \sin \alpha)/g$, the times of flight are

$$T_1 = (2 \times 50 \times \sin 26° 34')/10 = 4 \cdot 473 \, \text{s}$$

and $T_2 = (2 \times 50 \times \sin 63° 26')/10 = 8 \cdot 945 \, \text{s} .$

Thus $T_2 - T_1 = 4 \cdot 472 \, \text{s} = 4 \cdot 5 \, \text{s}$ to nearest tenth of a second.

We consider the maximum height reached by the ball in its flight. This will occur after a time $T/2 = (V \sin \alpha)/g$, and the corresponding height

$$H = V \sin \alpha \times \frac{V \sin \alpha}{g} - \frac{g}{2}\left(\frac{V \sin \alpha}{g}\right)^2$$

$$= (V^2 \sin^2 \alpha)/2g,$$

which is $\dfrac{50^2 \sin^2 26° 34'}{20}$ for the lower trajectory. Thus $H = 25$ m.

Since this is less than 30 m, the ball could not clear a tree 30 m high anywhere in its line of flight,   Q.E.D.

Question 32 :   (*G.C.E., Summer 1975, B.*)

A particle is projected with velocity V and elevation α from a point O. Show that the equation of the path of the particle, referred to horizontal and vertical axes Ox and Oy respectively in the plane of the path, is

$$y = x \tan \alpha - gx^2(1 + \tan^2 \alpha)/(2V^2).$$

A particle P is projected with velocity 70 m/s from the top of a vertical tower, of height 40 m, standing on a horizontal plane. The particle strikes the plane at a distance 200 m from the foot of the tower. Find the two possible angles of projection.

If these angles are α_1, α_2 where $\alpha_1 > \alpha_2$, calculate

(*a*) the greatest height correct to the nearest metre above the top of the tower when P is projected at inclination α_1,

(*b*) the time of flight, correct to the nearest tenth of a second, when P is projected at inclination α_2. (Take g as 9·8 m/s².)

Answer :

If (x, y) are the coordinates of the particle referred to Ox and Oy at time t, then $x = Vt \cos \alpha$, since the particle's horizontal velocity is constantly $V \cos \alpha$.

For the vertical motion, the initial velocity is $V \sin \alpha$, and so
$$y = Vt \sin \alpha - \tfrac{1}{2} gt^2.$$
Now, $t = x/(V \cos \alpha),$

$\therefore$ $y = \dfrac{Vx \sin \alpha}{V \cos \alpha} - \dfrac{g}{2}\left(\dfrac{x}{V \cos \alpha}\right)^2$

$$= x \tan \alpha - \frac{gx^2(1 + \tan^2 \alpha)}{2V^2},$$

since $\quad 1/\cos^2 \alpha = 1 + \tan^2 \alpha$, $\hfill$ Q.E.D.

From the data given, $y = -40$ when $x = 200$, and so

$$-40 = 200 \tan \alpha - \frac{9 \cdot 8 \times 40000 \,(1 + \tan^2 \alpha)}{2 \times 4900}$$

$$= 200 \tan \alpha - 40 - 40 \tan^2 \alpha.$$

Thus $\quad \tan^2 \alpha - 5 \tan \alpha = 0$,
giving $\quad \alpha = 0°$ or $\alpha = \tan^{-1} 5 = 78° \, 41'$.
So $\quad \alpha_1 = 78° \, 41'$ and $\alpha_2 = 0°$.

(*a*) The maximum height H above the point of projection is given by $H = (V^2 \sin^2 \alpha_1)/2g$, and so in this case

$$H = \frac{70^2 \sin^2 78° \, 41'}{2 \times 9 \cdot 8} = 240 \cdot 3 \text{ m}$$

$$= 240 \text{ m to the nearest metre.}$$

(*b*) For α_2, the particle is projected horizontally with velocity 70 ms^{-1}. Since this remains constant, the time t to cover 200 m is given by

$$t = \frac{200}{70} = 2 \cdot 9 \text{ s to the nearest tenth of a second.}$$

Section 11 : Forces in Equilibrium

[*For an assembly of particles in equilibrium under the action of a system of applied forces, (i) the vector sum of the forces is zero, and (ii) the sum of the moments of the forces about any point is zero. If the forces are co-planar, principle (i) may be used to give the sum of the components along any two independent (and usually perpendicular) directions as zero, while principle (ii) yields one equation relating the forces. In applying principle (i), any given force will not enter into the equation if the direction of resolution is taken perpendicular to that force; the same is true in applying principle (ii) if moments are taken about a point lying on the line of action of the given force.*

If three forces are in equilibrium, they must be both co-planar and concurrent. They may be represented by the sides of a triangle with sides parallel to the direction of the forces, and hence each force is proportional to the sine of the angle between the other two.

If a body P in contact with a surface T tends to move in a particular direction (towards the right in the diagram), a frictional force F in the opposite direction is produced. This can increase up to a maximum value, $F = \mu R$, where R is the normal reaction of T on P, and μ is the coefficient of friction for the surface. This situation is one of limiting friction, and here the total reaction S on P will make an angle $\lambda = \tan^{-1} \mu$ with the direction of R; λ is the angle of friction.

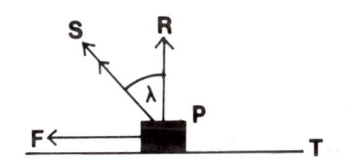

If an elastic string of natural length a is stretched to a length x, the tension T in it is given by

$$T = (x - a)\lambda/a ,$$

where λ is the modulus of elasticity of the string. The total work done in stretching the string from its initial length a to its final length x is

$$W = \int_a^x T\, dx = (x - a)^2 \lambda / 2a \;.]$$

Question 33 : (G.C.E., Specimen Paper for 1977, A.)

A uniform rod AB of length $2a$ and weight W rests in limiting equilibrium at an angle θ to the horizontal, with one end A on a rough horizontal floor and the other end B against a smooth vertical wall. The rod is in a vertical plane perpendicular to the wall. Show that the coefficient of friction between the rod and the floor is $\frac{1}{2} \cot \theta$.

Answer:

The rod is in equilibrium under the joint action of its weight W, the normal reaction R at B and the reaction S at A (which is the resultant of the normal reaction of the floor and the friction at A). These three forces must therefore be concurrent at Z. The coefficient of friction is given by $\mu = \tan \phi$.

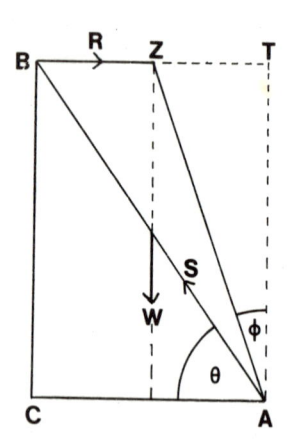

Therefore $$\mu = \frac{ZT}{AT} = \frac{1}{2} \cdot \frac{AC}{BC}$$
$$= \frac{1}{2} \cot \theta, \qquad Q.E.D.$$

Question 34 : (G.C.E., Summer 1975, A.)

A uniform rod PQ, of mass m and length $4a$, is smoothly pivoted to a fixed point at its end P, and the end Q is attached by a light spring, obeying Hooke's Law and of unstretched length $4a$, to a fixed point S at a distance $3a$ vertically above P. If the rod is in equilibrium in a horizontal position, find the tension in the spring and show that the force required to double its length is $10\, mg/3$.

FORCES IN EQUILIBRIUM

Answer :

Since the reaction at P is unknown, we take moments about P to find T. Thus

$2a \cdot mg = T \cdot 3a \sin \theta$,

whence $T = (2\ mg)/(3 \times 4/5)$

$= (5/6)\ mg$ (since $\sin \theta = 4/5$).

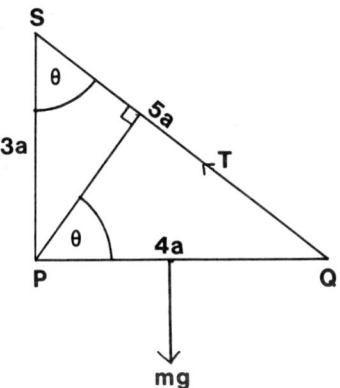

According to Hooke's Law, $T = \lambda \times \dfrac{extension}{original\ length}$,

where λ is the coefficient of elasticity for the spring, so that here

$$\frac{5\ mg}{6} = \lambda \cdot \frac{a}{4a}, \quad \text{i.e.} \quad \lambda = (10/3)\ mg.$$

If the length of the string is doubled, the extension equals the original length, and so the tension required is $\lambda = (10/3)\ mg$, Q.E.D.

Question 35 : (*G.C.E., Specimen Paper for 1977, B.*)

(*i*) Forces $\mathbf{i} + 3\mathbf{j}$, $-2\mathbf{i} - \mathbf{j}$, $\mathbf{i} - 2\mathbf{j}$ act through the points with position vectors $2\mathbf{i} + 5\mathbf{j}$, $4\mathbf{j}$, $-\mathbf{i} + \mathbf{j}$ respectively. Prove that this system of forces is equivalent to a couple and find its magnitude.

(*ii*) Three forces are represented in magnitude, direction and line of action by the sides AB, BC, CA of the triangle ABC. Show that this system of forces reduces to a couple and find the magnitude of this couple.

The force acting along the side CA is now reversed in direction. Find completely the resultant of this new system.

Answer:

(*i*) The resultant force of the given system is

$$(\mathbf{i} + 3\mathbf{j}) + (-2\mathbf{i} - \mathbf{j}) + (\mathbf{i} - 2\mathbf{j}) = \mathbf{O},$$

and so the system is equivalent to a couple.

The moment of the couple is found by calculating the sum of the moments of each of its constituent forces about any fixed point; for convenience we choose the origin.

For a force $\mathbf{F} = F_x\mathbf{i} + F_y\mathbf{j}$ acting through the point (X, Y), the moment of $\mathbf{F}$ about the origin is $YF_x - XF_y$ in a clockwise direction. Applying this formula to each of the given three forces and adding, we find that the moment of the couple is

$$5 \cdot 1 - 2 \cdot 3 + 4(-2) - 0(-1) + 1 \cdot 1 - (-1)(-2) = -10.$$

The moment of the couple thus has magnitude 10.

(*ii*) The resultant force

$$\mathbf{F} = \overrightarrow{AB} + \overrightarrow{BC} + \overrightarrow{CA} = \mathbf{O},$$

and so the system is equivalent to a couple. Taking moments about B to obtain the magnitude M of this couple gives

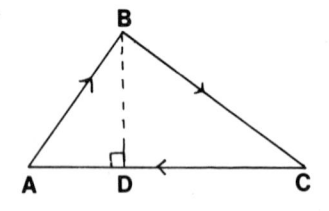

$$M = AC \times AD = 2 \times \text{area of } \triangle ABC.$$

If the force along CA is now reversed in direction, the new resultant is $\overrightarrow{AB} + \overrightarrow{BC} + \overrightarrow{AC}$. But $\overrightarrow{AB} + \overrightarrow{BC} = \overrightarrow{AC}$, so the new resultant is simply $2\overrightarrow{AC}$.

Consideration of moments about B shows that the line of action of this resultant passes through the mid-point of BD.

Question 36: (*G.C.E., Summer 1976, B.*)

Forces 2, 4, 6, 2*p*, 2*q* and 18 newtons act along the sides *AB*, *BC*, *CD*, *ED*, *EF* and *AF* respectively of a regular hexagon *ABCDEF*, the directions of the forces being indicated by the order of the letters. If the system is in equilibrium, find, by resolving parallel and perpendicular to AB, the values of p and q. Check your result by finding the moment of the forces about O, the centre of the hexagon.

The forces along ED, EF and AF are now replaced by a coplanar force through O and a coplanar couple. If the resulting system is in equilibrium and if the length of each side of the hexagon is 2 metres, calculate

(a) the magnitude of this force through O,

(b) the magnitude of the couple.

Answer :

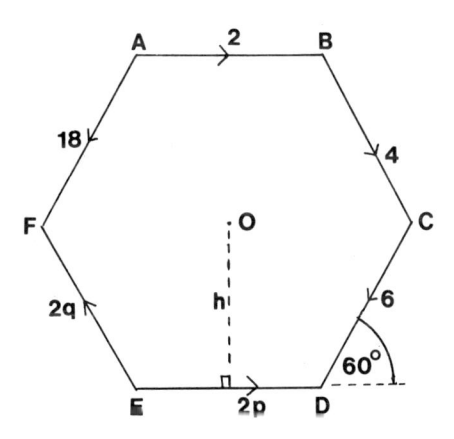

Resolving parallel to AB,
$$2 + 4\cos 60° - 6\cos 60° + 2p - 2q\cos 60° - 18\cos 60° = 0 ;$$
hence, since $\cos 60° = 1/2$,
$$2p - q = 8 . \tag{1}$$

Resolving perpendicular to AB,

$$4 \sin 60° + 6 \sin 60° - 2q \sin 60° + 18 \sin 60° = 0,$$

and so $q = 14$.

Substituting this into Equation (1) gives $p = 11$.

The perpendicular distance of each of the forces from the centre O of the hexagon is the same, viz. h. The net moment of all the forces about O is thus

$$h(2 + 4 + 6 - 2p + 2q - 18),$$

which is zero when the above values of p and q are inserted.

(*a*) The forces shown in the second diagram are in equilibrium with a coplanar force **F** through O and a coplanar couple M. The resultant of the three given forces is obtained by resolving parallel and perpendicular to AB. The resolute parallel to AB is

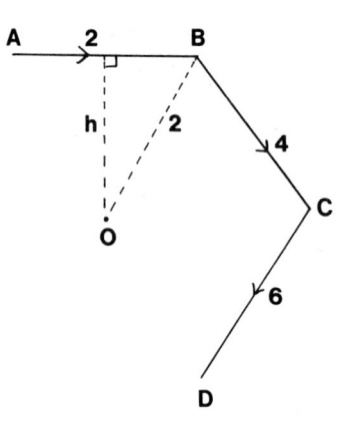

$$X = 2 + 4 \cos 60° - 6 \cos 60° = 1,$$

and that perpendicular to AB is

$$Y = 4 \sin 60° + 6 \sin 60° = 5\sqrt{3}.$$

Hence $\quad F = (X^2 + Y^2)^{1/2} = (1 + 75)^{1/2}$

$$= 2\sqrt{19} \text{ newtons}.$$

(b) Since $h = (4 - 1)^{1/2} = \sqrt{3}$ metres, the net moment of the given forces about O is
$$G = (2 + 4 + 6)\sqrt{3} = 12\sqrt{3},$$
and so $M = 12\sqrt{3}$ Nm.

Question 37 : (*G.C.E., Summer 1976, B.*)

A uniform rod AB of weight W and length $4a$ rests in a vertical plane with its end A on a rough horizontal plane and a point C of the rod, where $AC = 3a$, in contact with a smooth peg. If the rod makes an angle θ with the horizontal show that the force exerted by the peg on the rod is $\tfrac{2}{3}W \cos \theta$ and find, in terms of W and θ, the normal and frictional components of the force exerted by the plane on the rod at A. Deduce that for equilibrium to be possible, the coefficient of friction μ between the rod and the plane at A cannot be less than
$$\frac{\sin 2\theta}{2 - \cos 2\theta}.$$

Answer :

Let P and R be respectively the normal and frictional components of the reaction on the rod at A, and let S be the force exerted on the rod by the peg. Taking moments about A for the equilibrium of the rod gives
$$2aW \cos \theta = 3aS,$$
whence $S = \tfrac{2}{3}W \cos \theta$.

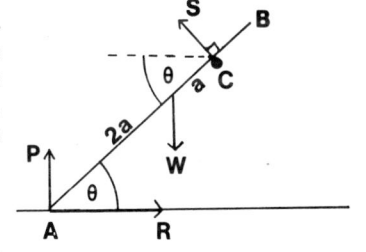

R can be found by resolving horizontally and P by resolving vertically:
$$R = S \sin \theta = \tfrac{2}{3}W \cos \theta \sin \theta,$$
and $P = W - S \cos \theta = W(1 - \tfrac{2}{3} \cos^2 \theta)$.

For equilibrium to be possible, it is necessary that
$$\mu \geqslant \frac{R}{P} = \frac{2 \cos \theta \sin \theta}{3 - 2 \cos^2 \theta} = \frac{\sin 2\theta}{2 - \cos 2\theta}, \quad Q.E.D.$$

Section 12 : Probability

[*In standard notation, $P(A \cap B) = P(B) \cdot P(A|B)$. If A and B are independent, $P(A|B) = P(A)$, and then $P(A \cap B) = P(A) \cdot P(B)$; also, $P(A \cup B) = P(A) + P(B)$.*

If the set A of possible events is $\{A_m\}$ and the set B is $\{B_n\}$, with $\sum_m P(A_m) = 1 = \sum_n P(B_n)$ and the A_i and B_j independent, and if a 'favourable' result corresponds to the pair (m, n) lying within a region T, then the probability of a favourable result is $\sum_{m,n} P(A_m) \cdot P(B_n)$, where the sum is taken over the region T.

If $P(x)$ is the probability that a variable equals x, then the expected value of the variable is $\sum_x xP(x)$, where the sum is taken over all possible values of x.]

Question 38 : (*G.C.E., Specimen Paper for 1977, A.*)

A red die and a blue die are rolled and the sum of their scores is called the *tot*.

(*a*) What is the probability of scoring a *tot* of 8 or more?

(*b*) What is the probability of scoring a *tot* of 8 or more given that the red die score is 4?

Answer :

(*a*) If the red die scores 6, the *tot* will be 8 or more provided the blue die scores 2, 3, 4, 5 or 6; that is, if it scores one of five possibilities. For a score of 5 on the red die there are four possibilities on the blue: 3, 4, 5, 6. Similarly for scores of 4, 3 and 2 on the red die, the number of possibilities on the blue are 3, 2 and 1 respectively. Thus the total number of combinations of the two scores leading to a *tot* of 8 or more is $5 + 4 + 3 + 2 + 1 = 15$. But the total number of possible combinations of the scores on the two dice is $6 \times 6 = 36$. Thus the probability of a *tot* of 8 or more is $15/36 = 5/12$.

(*b*) If the red die score is 4, the *tot* will be 8 or more only if the blue die scores 4, 5 or 6, giving three possibilities. Since the total number of possible scores on the blue die is six, the probability of scoring a *tot* of 8 or more is $3/6 = 1/2$.

PROBABILITY

Question 39 : (*G.C.E., Specimen Paper for 1977, B.*)

At a shooting gallery, each time a marksman fires the probability of his hitting a clay pipe is $\frac{1}{4}$.

What is the probability of his first hitting a pipe

(*a*) in four shots or less,

(*b*) with his third shot?

The marksman pays 3p for each shot and gets a reward of 15p when he hits a pipe. Show that he may expect to win 3p if he plays the game until he hits a pipe.

(You may assume that $1 + 2x + 3x^2 + \ldots = (1-x)^{-2}$.)

Answer :

P (First hit occurs on nth shot)

$= P$ (First $(n-1)$ shots miss and nth shot hits).

As each shot is assumed to be independent of the others, this is

P (First $(n-1)$ shots miss) . P(nth shot hits)

$$= \left(\frac{3}{4}\right)^{n-1} \times \frac{1}{4} = \frac{3^{n-1}}{4^n}.$$

(*a*) Thus the probability of first hitting a pipe in 4 shots or fewer, which is the sum of the probabilities of first hitting in 1, 2, 3 and 4 shots, is

$$\frac{1}{4}\left[1 + \frac{3}{4} + \left(\frac{3}{4}\right)^2 + \left(\frac{3}{4}\right)^3\right] = \frac{175}{256}.$$

(*b*) The probability of first hitting a pipe with the third shot is, as shown above,

$$\left(\frac{1}{4}\right)\left(\frac{3}{4}\right)^2 = \frac{9}{64}.$$

Since the marksman pays 3p per shot and receives 15p when he hits a pipe, he will win $(15 - 3n)$p if his first hit is with the nth shot. The probability of this occurring is $(3/4)^{n-1}(1/4)$, as shown above, and so the amount he would expect to win is

$$X = \sum_{n=1}^{\infty} (15 - 3n)(1/4)(3/4)^{n-1}$$

$$= 15 \sum_{n=1}^{\infty} (1/4)(3/4)^{n-1} - (3/4) \sum_{n=1}^{\infty} n(3/4)^{n-1}.$$

$\sum_{n=1}^{\infty} (1/4)(3/4)^{n-1} = 1$ since it is the sum of the probabilities of all possible outcomes, and it is given that $\sum_{n=1}^{\infty} nx^{n-1} = (1-x)^{-2}$.

Therefore
$$X = 15 - \frac{3}{4}\left[1 - \frac{3}{4}\right]^{-2}$$

$$= 15 - \frac{3}{4} \cdot 16 = 3.$$

The marksman may therefore expect to win 3p,   Q.E.D.

Multiple-Choice Questions

[*The following section illustrates the various types of question that may be expected. All examples are chosen from the Specimen Paper 1 for 1977 of the University of London, who require 30 questions to be answered in 1 hour. This means that you have on average two minutes for each question, and it is therefore most important not to linger over a question you find difficult. Answers (with brief notes on the reasoning) are given at the end of this section. All correct answers carry equal marks.*

The only materials required for Paper 1 are an HB pencil and a soft rubber. Tables, slide-rules and drawing instruments are not to be used for this paper.]

Directions *for QQ. 40, 41 and 42:* each of these questions is followed by five suggested answers. Select the correct answer in each case.

Question 40:

A particle P moves along a straight line Ox so that, at time t, $x = a \cos \omega t + a \sin \omega t$, where $OP = x$ and a and ω are constants. The acceleration of P at that instant is given by

A $a \cos \dfrac{\omega}{t} + a \sin \dfrac{\omega}{t}$ **D** $-a\omega^2 \cos \omega t - a\omega^2 \sin \omega t$

B $\dfrac{a}{t^2} \cos \omega t + \dfrac{a}{t^2} \sin \omega t$ **E** $a\omega^2 \cos \omega t + a\omega^2 \sin \omega t$

C $a\omega \cos \omega t - a\omega \sin \omega t$

Question 41:

$P = \dfrac{GM_1 M_2}{D^2}$ is a relation involving a force P, masses M_1 and M_2, a displacement D and a constant quantity G. What are the units of G in the SI system?

A G is dimensionless   **C** $kg^{-1}m^3s^{-1}$ **E** $kg^{-2}m^2$

B $kg^{-1}m^3s^{-2}$ **D** $kg^{-1}m^2$

Question 42 :

If the likelihood of giving birth to a boy equals the likelihood of giving birth to a girl, what is the probability that in a family of 3 children all are girls?

A $\frac{1}{2}$ C $\frac{1}{3}$ E $\frac{1}{8}$

B $\frac{3}{8}$ D $\frac{1}{4}$

Directions *for QQ. 43, 44 and 45:* for each of the following questions, one or more of the responses given are correct. Decide which of the responses is (are) correct. Then choose

A   if **1**, **2** and **3** are correct   D   if only **1** is correct

B   if only **1** and **2** are correct   E   if only **3** is correct

C   if only **2** and **3** are correct

Question 43 :

With the usual notation, the acceleration of a particle moving in a straight line is given by

1   $\dfrac{d^2x}{dt^2}$ 2 $\dfrac{v\,dv}{dx}$ 3 $\dfrac{dv}{dt}$

Question 44 :

Two perfectly elastic spheres of equal mass, moving towards each other along the same straight line with equal speeds, collide.

1   The total linear momentum remains constant

2   The total kinetic energy remains constant

3   The two spheres are brought to rest

MULTIPLE-CHOICE QUESTIONS

Question 45 :

Three forces are represented in magnitude, direction and line of action by $\vec{PQ}$, $\vec{RQ}$ and $\vec{RP}$. The resultant of these three forces

1 is equal in magnitude to $2\,\vec{QR}$

2 passes through the mid-points of PQ and PR

3 is in the direction $\vec{QR}$

Directions *for QQ. 46 and 47:* each of the following questions consists of two statements. Determine the relationship between these statements and choose

A if **1** always implies **2** but **2** does not imply **1**

B if **2** always implies **1** but **1** does not imply **2**

C if **1** always implies **2** and **2** always implies **1**

D if **1** always denies **2** and **2** always denies **1**

E if none of the above relationships holds

Question 46 :

A particle P moves along a straight line Ox, and $OP = x$.

1 At time t, the velocity v of the particle is given by $v^2 = kx^4$

2 At time t, the acceleration of the particle is $4kx^3$

Question 47 :

1 $\mathbf{a} = \mathbf{b}$ \qquad 2 $|\mathbf{a}| = |\mathbf{b}|$

Directions for *QQ. 48 and 49:* each of the following questions consists of a problem followed by four pieces of information. Do not actually solve the problem, but decide whether the problem could be solved if any of the pieces of information were omitted, and choose

A   if **1** could be omitted   D   if **4** could be omitted

B   if **2** could be omitted   E   if none of them could be omitted

C   if **3** could be omitted

Question 48 :

A heavy particle is attached to one end of a light elastic string obeying Hooke's law, and the other end of the string is tied at a fixed point O. The particle is then set in motion under gravity. Find the depth of the particle below O when it first comes instantaneously to rest.

1   The unstretched length of the string is l m

2   The mass of the particle is m kg

3   The modulus of elasticity of the string is λ N

4   The particle is dropped from rest at O

Question 49 :

A projectile is fired from ground level towards a wall 10 m high. Determine whether the projectile will pass over the wall.

1   The initial horizontal component of the velocity of the projectile is 20 m/s

2   The initial vertical component of the velocity of the projectile is 20 m/s

3   The horizontal distance of the wall from the point of projection is 30 m

4   The mass of the projectile is 10 grams

Answers to Multiple-Choice Questions

Q.40 : D

[Acceleration is d^2x/dt^2.]

Q.41 : B

[Expressing the given equation in dimensions,

$$\frac{[M][L]}{[T]^2} = \frac{[G][M]^2}{[L]^2}, \text{ so } [G] = \frac{[L]^3}{[M][T]^2}.$$

Hence the units of G are $m^3 kg^{-1} s^{-2}$.]

Q.42 : E

[Since $P(B) = P(G)$ and $P(B) + P(G) = 1, P(G) = 1/2$. Hence the probability that all 3 children are girls is $(1/2)^3 = 1/8$.]

Q.43 : A

[Acceleration $a = dv/dt$, and $v = dx/dt$. Thus $a = d^2x/dt^2$. Also $\frac{dv}{dt} = \frac{dv}{dx} \times \frac{dx}{dt} = \frac{dv}{dx} \times v$.

Hence $a = v \frac{dv}{dx}$.]

Q.44 : B

[**(1)** is true, since this always holds for any type of collision;

(2) is true since the spheres are perfectly elastic;

(3) is false since the total K.E., which remains constant, is non-zero.]

Q.45 : B

[$\vec{RP} + \vec{PQ} = \vec{RQ}$, and so the resultant of the 3 forces is $\vec{RQ} + \vec{RQ} = 2\vec{RQ} \equiv \mathbf{Z}$.

(1) is true since $|\mathbf{Z}| = 2\,QR$;

(2) is true since taking moments about P shows that the distance of the resultant from P is half that of the side QR from P;

(3) is false since $\mathbf{Z}$ is in the direction $\overrightarrow{RQ}$.]

Q.46 : D

[Since $a = v\,dv/dx = \tfrac{1}{2}d(v^2)/dx$, $a = 2kx^3$ if $v^2 = kx^4$. If $a = 4kx^3$, $v^2 = 2kx^4 + C$.]

Q.47 : A

[If two vectors are equal, their magnitudes must be equal. However, if their magnitudes are equal, the vectors may still be unequal since they may have different directions.]

Q.48 : E

[The force on the particle when it performs simple harmonic motion clearly depends on λ and l, and its acceleration thus depends on λ, l and m. Its depth below O when at rest will depend on its acceleration and on the initial conditions, and therefore all four pieces of information are needed to solve the problem.]

Q.49 : D

[The trajectory of a projectile for a given initial velocity is independent of the projectile's mass, and thus **4** can be omitted.]

MODEL TEST PAPER INDEX

Like many other Examination Boards, London University have ceased to treat Applied Mathematics as a separate subject, although most questions on it are still grouped together in one of the papers. The new, combined Syllabus D provides for a Multiple-Choice Paper, and two further papers each consisting of 8 short questions (of which all are to be attempted) and 8 harder questions (of which 4 are to be attempted). Total subject marks are allocated approximately as follows:

MULTIPLE-CHOICE	20%
SHORT QUESTIONS	28% (i.e. $3\frac{1}{2}$% each)
HARDER QUESTIONS	52% (i.e. 13% each).

Test papers equivalent to the Applied Mathematics element of the combined syllabus may be constructed by selecting the following questions from this book. Full marks may be obtained for answers to all the questions in Section A and to four questions chosen from Section B. If more than four questions in Section B are attempted, only the best four answers will be taken into account.

PAPER I :

    Section A : QQ. 1, 5, 7, 10, 15, 19, 28, 38;

    Section B : QQ. 3, 9, 13, 16, 25, 30, 35, 36.

                Time allowed: $2\frac{1}{2}$ hours.

PAPER II :

    Section A : QQ. 6, 8, 11, 20, 24, 29, 33, 34;

    Section B : QQ. 2, 4, 14, 18, 22, 26, 31, 39.

                Time allowed: $2\frac{1}{2}$ hours.

A note on the Multiple-Choice Paper, with examples, will be found on pp. 57–62.